YOUR KNOWLEDGE HAS VALUE

- We will publish your bachelor's and
  master's thesis, essays and papers

- Your own eBook and book -
  sold worldwide in all relevant shops

- Earn money with each sale

Upload your text at www.GRIN.com
and publish for free

Characterization of Resonant Coupled Inductor in a Wireless Power Transfer System

Alan Nebrida

Bibliographic information published by the German National Library:

The German National Library lists this publication in the National Bibliography; detailed bibliographic data are available on the Internet at http://dnb.dnb.de.

ISBN: 9783346718303
This book is also available as an ebook.

© GRIN Publishing GmbH
Nymphenburger Straße 86
80636 München

Print and binding: Books on Demand GmbH, Norderstedt, Germany
Printed on acid-free paper from responsible sources.

The present work has been carefully prepared. Nevertheless, authors and publishers do not incur liability for the correctness of information, notes, links and advice as well as any printing errors.

GRIN web shop: https://www.grin.com/document/1271343

CHARACTERIZATION OF RESONANT COUPLED INDUCTOR IN A WIRELESS POWER TRANSFER SYSTEM

Alan P. Nebrida – Nueva Vizcaya State University

Abstract

Wireless power transfer based on coupled magnetic resonances is a new technology in which energy can be transferred via coupled magnetic resonances in the non-radiative near-field. This paper presents the design, simulation, fabrication, and experimental characterization of a single-loop inductor that acts as the receiver and transmitter of the system. A circuit model is presented to provide a convenient reference for analysis of the transfer characteristics of a magnetically coupled resonator system. Based on this structure, the output voltage in the receiving loop is related with different transfer distances and orientations. At a given driving frequency was simulated and analysed. The driving resonant frequency of the system is approximately 580 kHz. Experimental results show that energy can still be transferred even the receiver is sheltered in most conditions. Non-metallic objects such as walls, books, wooden products, organic glass panels, leather, and textiles have no impact on power transfer. Energy transfer shows that transfer efficiency varies inversely as the square of the distance $\left(\frac{1}{r^2}\right)$ between the transmitting and receiving resonance loops. The longer is the distance between them, the lower the transfer power and efficiency is. This depicts the on near field theory.

Keywords: Wireless Power Transfer, Coupled Magnetic Resonance, Resonance Frequency, Inductive Coupling

Introduction

Nowadays, the manufacturing and improvement of mobile appliances such as mobile phones, laptops and other electronics and communications devices have increased rapidly. Regardless of the portability and mobility to communicate wirelessly these devices require regular charging usually by plugging them into a wall outlet. Thus, intensifies the quest for new techniques in order to provide power wirelessly to these devices to improve its portability and mobility for end users.

Wireless energy transfer or Wireless Power Transfer (WPT) is a process that takes place in a system where electrical energy is transmitted from a power source to an electrical load without interconnecting wires. At present, energy has been transferred wirelessly using diverse physical mechanisms like: laser, Piezoelectric Principle, radio waves and Microwaves, Inductive Coupling and Strong Electromagnetic Resonance. Transferring great quantities of power using magnetic field inevitably creates unrest and poses health hazards to humans. Although the method is efficient, it has a disadvantage since it requires a line of sight and it is a dangerous mechanism for living beings. Thus, wireless energy transfer using phenomenon of electromagnetic resonance has become a viable option, at least for short distances, since it has high efficiency for power transfer and does not affect human health. (Karalis et al., 2008)

The technology of WPT, electromagnetic induction and microwave power transfer are famous. However, electromagnetic resonance couplings have only been proposed recently. The technology of this wireless power transfers requires three main elements: large air gaps, high efficiency and a large amount of power. Electromagnetic resonance coupling is the only technology that deals with these three elements.

The general objective of this study is to characterize resonant coupled inductor for wireless power transfer system. Specifically, this study aims to: (1) Design a wireless power transfer system; (2) Create a model of a wireless power transfer system and simulate the system; (3) Determine the effective distance that the model can transmit for a given amount of power for different orientations of the power transmitter; (4) Compare the simulation results with measured values; and (5) Develop procedure to analyze data to come up with a method for characterizing wireless power transmission systems.

In the development of household devices, engineers and designers today are expected to provide a new level of convenience and flexibility to consumers. A WPT system was investigated to overcome the inconvenience of using power cable. Wireless transmission is useful in cases where interconnecting wires are inconvenient, hazardous, or impossible. When wireless power transfer is achieved, the portability and mobility of electronic devices will improve because the process of charging devices will be made a lot more convenient as we do not have to plug a cord into a socket. Also, in this case of wireless charging, the danger of being electrocuted due to wear and tear of an old cord makes the charging safe.

This research project use resonant inductive coupling to transfer power wirelessly. The research study uses a low power supply to transmit power. The scope of this study is limited to the construction of a simplified WPT system using a resonant coupled inductor system. This study includes the matching sections, derivation of relationship between the coupling coefficient and distance and the parameters (quality factor, coupling coefficients, mutual inductance, resonance frequency) of the resonators. The researcher uses a 12V, 5W CYD LED bulb as the load to be able to distinguish easily whether the system is operating well or not.

This study will not cover other possible methods in improving the efficiency of a wireless power.

Methodology

This section outlines the design, approach and techniques followed in prototyping and testing of the proposed system. Shown in Figure 3.1 are the procedural steps used in conducting this research study.

As gleaned from the Methodology Flow Chart, data gathering and analysis covered the required definition of the hardware and software to come out with the appropriate design and the development of the project.

The design refers to the actual planning for the hardware and software of the system. It includes the determination of the conceptual framework and materials used for the hardware implementation. It also refers to the determination of algorithm, program flow and the type of software used for the software implementation.

After the project was assembled, testing and debugging followed. Al the necessary adjustment is already done. Also, the hardware and the software were integrated for testing.

Experiments were done considering the variables of the study which are the distances between the transmitting and receiving coils, resonant frequency, voltage gain and the efficiency of the system.

Characterization of the parameter of Resonant Coupled Inductor in a wireless Power Transfer System was based on the results of the experiment which are discussed in the following sections of this paper.

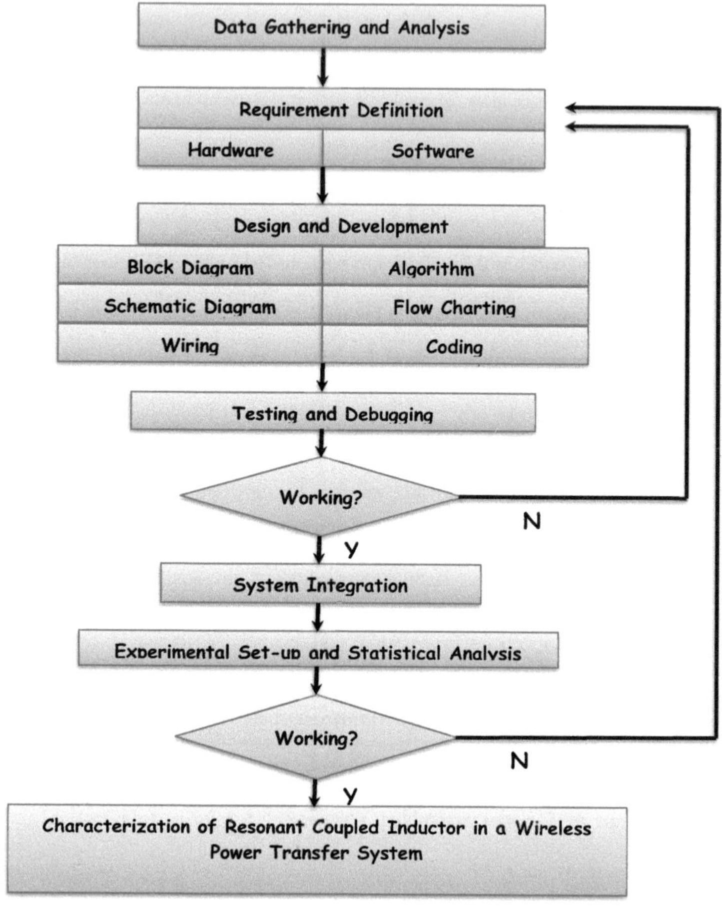

Figure 3.1. Methodology Flow Chart

Conceptualization/Development of System Design

System Block Diagram

Figure 3.2 shows the block diagram of the system. The system is composed of four main sections: the power amplifier; the transmitting loops; the receiving loops and the voltage rectifier, this systems are being design to modify the objectives.

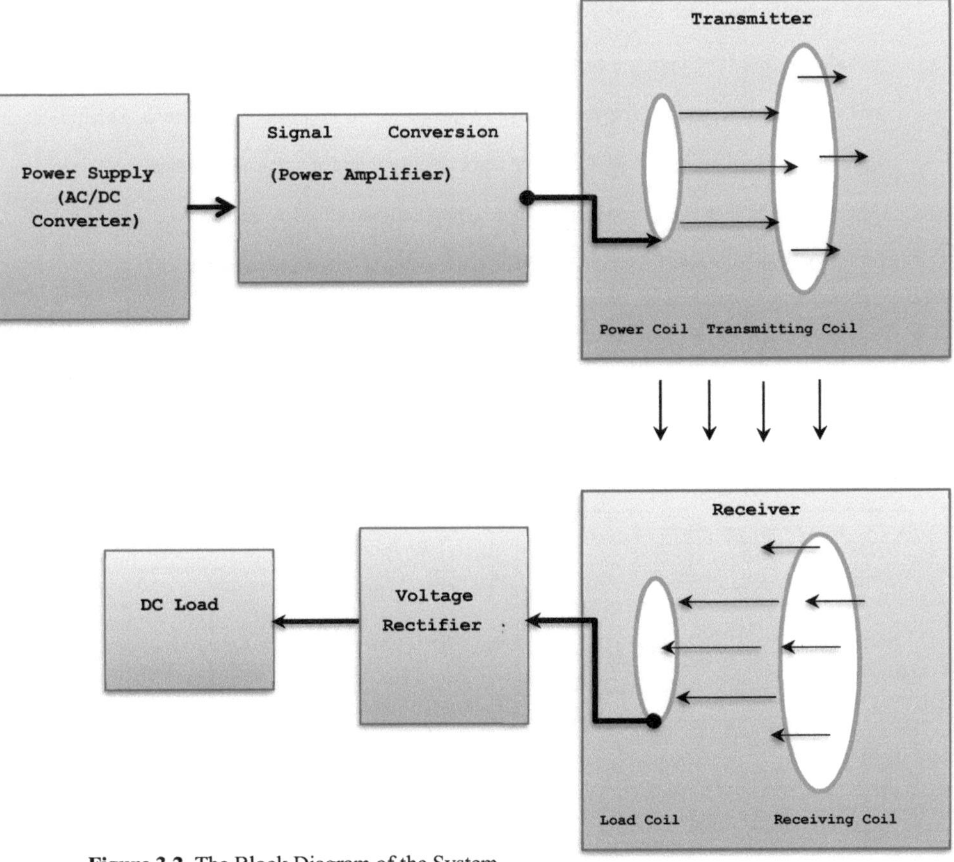

Figure 3.2. The Block Diagram of the System

Power Supply

The power supply of the transmitting circuit consists of 2 main sections: an AC/AC step-down transformer and a rectifier. The system was designed in accordance with the requirements of the amplifier circuit. The goal of the power supply is to convert AC wall outlet voltage to a steady DC voltage source. The transformer must reduce wall voltage down to a more manageable level in order to use the components in the amplifier. This transformer provides sufficient voltage and current for the rectifier. The rectifier transforms the AC signal into a DC signal so that the frequency can be adjusted via the amplifier and antenna. The full-wave bridge rectifier is the KBU4D which can be easily found at any Radioshack store. Large capacitors were then connected to the output of the full-wave bridge rectifier to ensure that a steady DC voltage could be maintained. The power supply schematic shown in Figure 3.3.

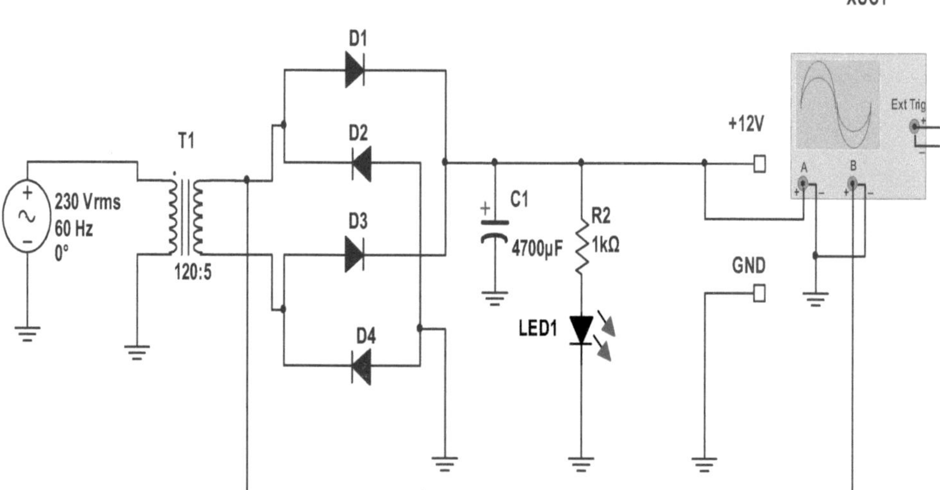

Figure 3.3. Schematic Diagram of Power Supply

Multisim was used to simulate the operation of the circuit. Figure 3.4: Simulation Result for the Power Supply shows the overall output of the system. The figure displays the voltage output before the rectifier (red line) and the voltage at the output (green line) is +12V, which meets the specification.

Although no tough design challenges were present in creating the power supply, it was necessary that the system operate well. The key points in creating a DC power supply are the voltage, current, and removing ripple in the DC components. All three of these key points were known and addressed in the design process.

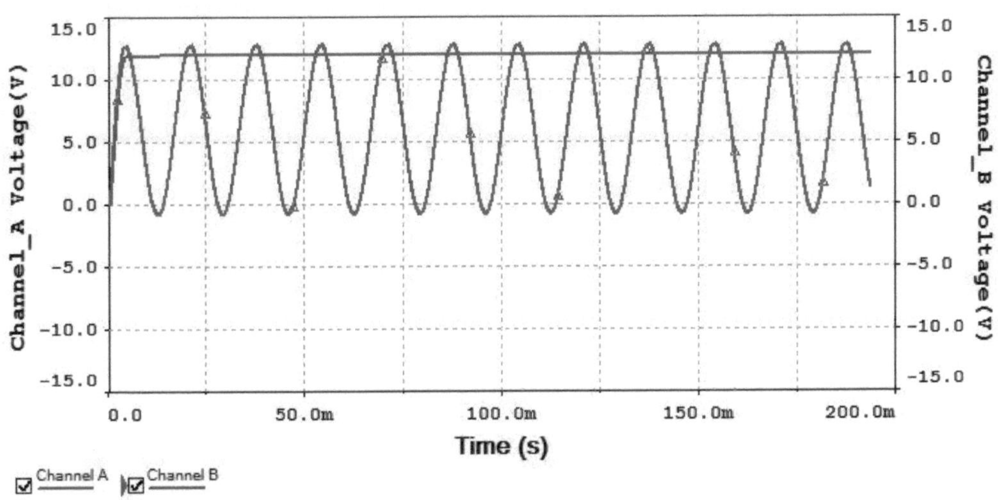

Figure 3.4. Simulation Result for the Power Supply

Amplifier

In order to generate the maximum amount of flux which will induce the largest voltage on a receiving loop, a large amount of current must be transferred into the transmitting loop. The power amplifier is capable of producing the necessary current. The goal of the amplifier is to take an oscillating signal and increase the voltage, while providing current to the transmitting antenna. The circuit oscillation is provided by a feedback loop from the antenna. Operating the circuit in this manner will allow the resonant frequency of the antenna to be fed back into the amplifier, which will then be amplified and transmitted.

Circuit simulations have been performed using Multisim. This software tool includes a library of parts that do not match the parts used in bench testing. Generic parts were used in their place for simulation purposes. The simulated circuit's layout is shown below in Figure 3.5: Schematic Diagram of Amplifier. The tank circuit, C_1 and L_3, represents the antenna load and the values were chosen to resonate at 580kHz. Also, Table 3, shows the different parameters used in the simulation with their respective values.

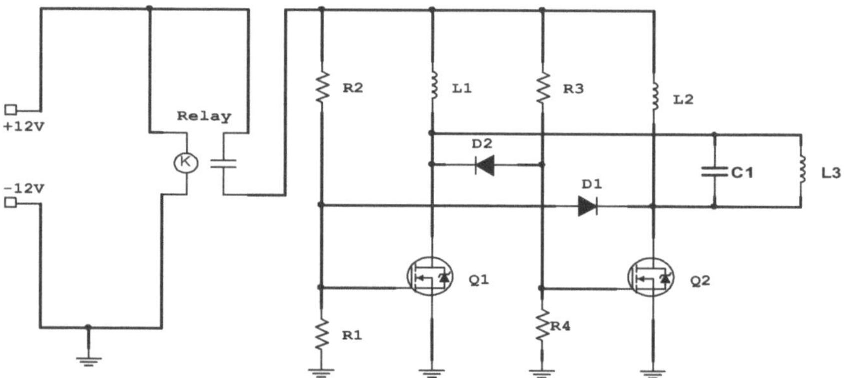

Figure 3.5. Schematic Diagram of Amplifier

Through this circuit, the theory of operation was verified. The tank circuit oscillated, creating the necessary voltage. This signal was then amplified through the mosfet. The output shown in (b): Amplifier Simulation Output, the start-up time of the test is approximately $3.5ms$ before the input signal reaches 80% of its final value. The use of generic components in the simulation limits the test, and Figure 3.6 (a) shows that only 581 kHz was achieved and approximately 36.7 volts peak to peak on the output.

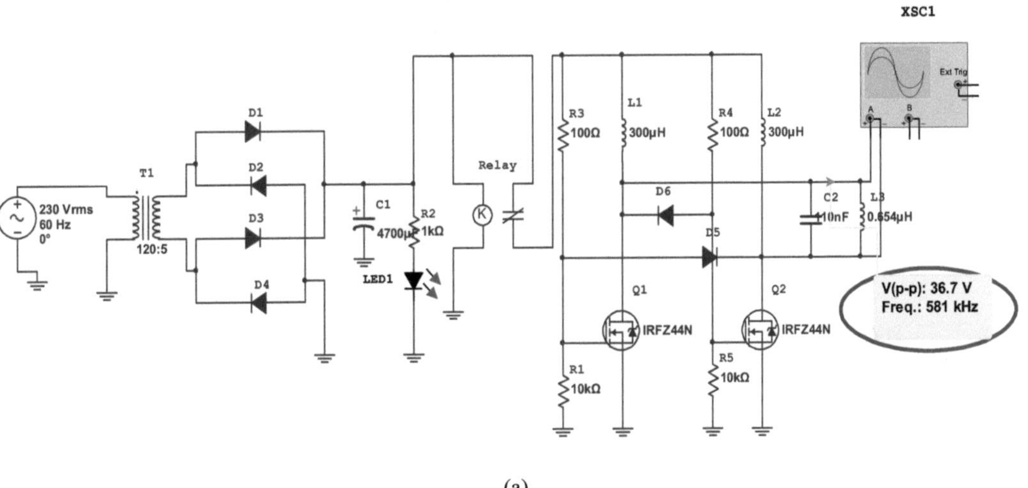

(a)

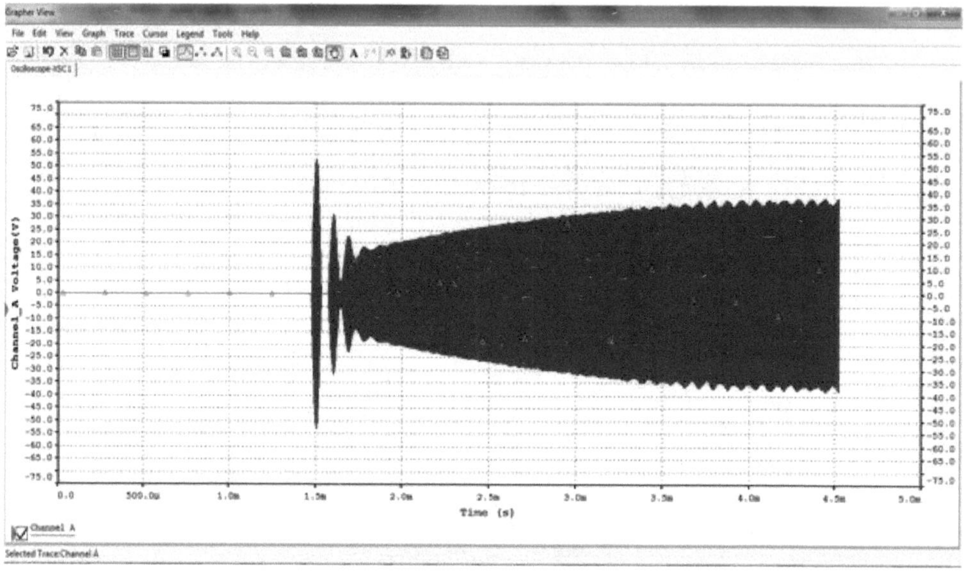

(b)

Figure 3.6. Amplifier output in Multisim using (a) Measurement Probe, (b) Oscilloscope

Transmitter and Receiver Loops

The transmitter and receiver circuits combined are called the coupling circuit. It is the heart of the entire system as the actual wireless power transfer happens out here. The efficiency of the coupling circuit is based on the separation distances between them. The separation between the coils was adjustable along their axis, thus measurements were performed across a dc load for several different distances, with the results plotted in Figure 3.13.

Voltage Rectifier

A rectifier is needed to rectify the AC voltage received from the load coil to drive a DC load. A type of circuit that produces an output waveform that generates an output voltage which is purely DC or has some specified DC component is a Full Wave Bridge Rectifier. This type of single phase rectifier uses four individual rectifying diodes connected in a closed loop "bridge" configuration to produce the desired output. The smoothing capacitor connected to the bridge circuit converts the full-wave rippled output of the rectifier into a smooth DC output voltage. Since the diodes had to rectify AC signals of Megahertz frequencies, fast signal diodes,31DF4, had to be used for the bridge circuit.

Circuit Model and Transfer System

The magnetically coupled resonator system can be represented in terms of lumped circuit elements L, C, and R. Figure 3.7 shows a circuit diagram that can be used for hand analysis or for SPICE simulations.

The schematic diagram in figure 3.7 consists of four resonant circuits, linked magnetically by coupling coefficients k_{12}, k_{23}, and k_{34}. Starting from the left, the power loop is excited by a source with finite output impedance, R_S. A simple one-turn power loop can be modelled as an inductor L_1 with parasitic resistance, R_1. A capacitor C_1 is added to make the power loop resonant at the frequency of interest. The transmitter (Tx) loop also consist of one-turn air core inductor L_2, with parasitic resistance R_2. The geometry of the Tx loop determines its self-capacitance which is represented as C_2. Inductors L_1 and L_2 are connected with coupling coefficient k_{12}; the receive side is defined similarly. Finally, the transmitter and receiver loops are linked by coupling coefficient, k_{23}. A typical implementation of the system would have the drive loop and Tx coil built into a single device such that k_{12} would be fixed. Similarly, k_{34} would also be fixed. Thus k_{23} is the remaining uncontrolled value which varies as a function of the distances between the transmitter and the receiver.

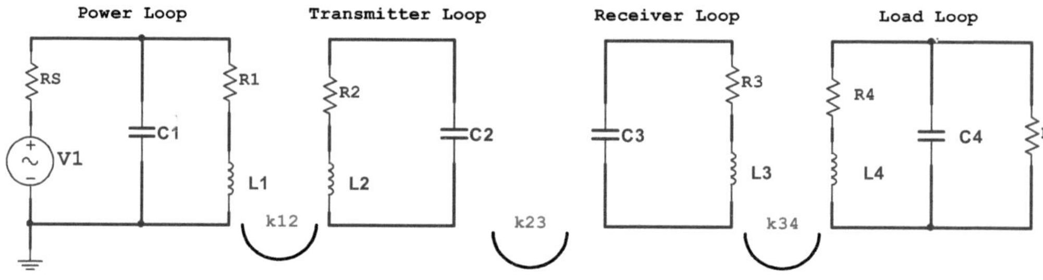

Figure 3.7:.Equivalent circuit model of the wireless power transfer system.

Each of the four antenna elements are modelled as parallel resonators, which are linked by mutual inductances and coupling coefficients.

The circuit model provides a convenient reference for analysis of the transfer characteristics of a magnetically coupled resonator system. For the sake of simplicity the cross-coupling term k_{12} and k_{34} are neglected in the following analysis. The circuit model offers a convenient way to systematically analyse the characteristics of the system. By applying circuit theory Kirchhoff's Voltage Law (KVL) to this system, with the currents in each resonant circuit chosen as illustrated in Figure 3.7, a relationship between current through each coil and the voltage applied to the power coil can be captured as in the following matrix in equation **3. 1**, where the coupling coefficient is defined in equation **3. 2.**

$$
\begin{bmatrix} V_s \\ 0 \\ 0 \\ 0 \end{bmatrix} = \begin{bmatrix} Z_1 & j\omega M_{12} & 0 & 0 \\ j\omega M_{12} & Z_2 & -j\omega M_{23} & 0 \\ 0 & -j\omega M_{23} & Z_3 & j\omega M_{34} \\ 0 & 0 & j\omega M_{34} & Z_4 \end{bmatrix} \begin{bmatrix} i_1 \\ i_2 \\ i_3 \\ i_4 \end{bmatrix} \tag{3.1}
$$

$$
k_{xy} = \frac{M_{xy}}{\sqrt{L_x L_y}}, \quad 0 \le k_{xy} \le 1 \tag{3.2}
$$

where M_{xy} is mutual inductance between coil "x" and coil "y" and Z_1, Z_2, Z_3 and Z_4 are

loop impedances of the four coils. These impedances can be indicated as below;

$$
Z_1 = R_s + \frac{R_1 + j\omega L_1}{\omega^2 L_1 - j\omega C_1 R_1 + 1} \tag{3.3}
$$

$$
Z_2 = R_2 + j\left(\omega L_2 - \frac{1}{\omega C_2}\right) \tag{3.4}
$$

$$
Z_3 = R_3 + j\left(\omega L_3 - \frac{1}{\omega C_3}\right) \tag{3.5}
$$

$$
Z_4 = R_L + \frac{R_4 + j\omega L_4}{\omega^2 L_4 - j\omega C_4 R_4 + 1} \tag{3.6}
$$

From the matrix **3.1**, by using the substitution method, the current in the load coil resonant

circuit is derived as

$$
i_4 = -\frac{j\omega^3 M_{12} M_{23} M_{34} V_s}{Z_1 Z_2 Z_3 Z_4 + \omega^2 M_{12}{}^2 Z_3 Z_4 + \omega^2 M_{23}{}^2 Z_1 Z_4 + \omega^2 M_{34}{}^2 Z_1 Z_2 + \omega^4 M_{12}{}^2 M_{34}{}^2} \tag{3.7}
$$

It is seen that the voltage across the load is equal to $V_L = -i_4 R_L$ and the relationship

between the voltages of source and load is given as V_L/V_s.

The system model can be considered as a two port network. To analyze a figure of

merit of this kind of system, S – parameter is a suitable candidate. Actually, S_{21} is a vector

referring to a ratio of signal exiting at an output port to a signal incident at an input port. This

parameter is important because a power gain, the critical factor determining power transfer efficiency, is given by $|S_{21}|^2$, the squared magnitude of S_{21}. The parameter of S_{21} is calculated by (Sample et al., 2011, as cited in Fletcher & Rossing, 1998; Mongia, 2007)

$$S_{21} = 2\frac{V_L}{V_S}\left(\frac{R_s}{R_L}\right)^{1/2} \tag{3.8}$$

Thus, combining with $M_{xy} = k_{xy}\sqrt{L_xL_y}$ derived from 3.2, the S_{21} parameter is given as

$$S_{21} = \frac{j2\omega^3 k_{12}k_{23}k_{34}L_2L_3\sqrt{L_1L_4R_sR_L}}{Z_1Z_2Z_3Z_4 + k_{12}^2L_1L_2Z_3Z_4\omega^2 + k_{23}^2L_2L_3Z_1Z_4\omega^2 + k_{34}^2L_3L_4Z_1Z_2\omega^2 + k_{12}^2k_{34}^2L_1L_2Z_3Z_4\omega^2}$$

$$\tag{3.9}$$

It is helpful to analyse the performance of the system according to equation 3.9. With all the circuit parameters provided in Table 3.1, the parameter regarded as the factor determining the efficiency of the system, magnitude of S_{21}, can be performed by a function of only two variables k_{23} and frequency. As referred to earlier, the coupling coefficient k_{23} is the parameter which varies according to changes in circumstances. A changeable distance, for instance, is a cause of k_{23} variation. In addition, changes in the orientation or misalignment between the transmitting and receiving resonators affect the above coefficient as well. Actually, when the distance increases, k_{23} will decrease because the mutual inductance between those coils declines with distance. In case of a variable orientation or misalignment, the k_{23} also changes. The relationships between S_{21}, k_{23} and frequency is demonstrated in Figure 3.8. From the figure shown, it is seen that when k_{23} is small in cases of the large distance between the transmitter and the receiver or the misalignment plus the orientation deviation taking place; the efficiency represented as S_{21} magnitude is able to reach a peak at the self-resonant frequency of approximately $580kHz$. Thus, the resonant frequency alters as k_{23} changes.

Table 3.1. Component values in the Circuit model.

Parameter	Value
R_S, R_L	50Ω
L_1, L_4	$0.654\mu H$
C_1, C_4	$110nF$
R_1, R_4	0.20Ω
L_2, L_3	$0.944\mu H$
C_2, C_3	$80nF$
R_2, R_3	0.20Ω
k_{23}	$0.001\ to\ 0.30$
f_o	$580\ kHz$
$frequency$	$100kHz\ to\ 1.5MHz$

It is instructive to analyse carefully a trend of $|S_{21}|$ as a function of k_{23}. When the coefficient k_{23} is very small corresponding to a case when the transmitter and the receiver are too far away from each other, $|S_{21}|$ is low. When the distance between the resonators is decreases, k_{23} increases bringing about a higher magnitude of $|S_{21}|$. However, as $|S_{21}|$ increases to a certain level, the higher k_{23} does not lead to the higher amount of $|S_{21}|$. Moreover, there is a frequency splitting issue which substantially reduces the system efficiency. The point, at which the deviation of the original resonant frequency (593.383 kHz) happens, plays a prominent role in the system. It clarifies the relative position of the resonators that the performance of the system is the highest. If the distance is longer than that range, the efficiency is poorly defined. On the contrary, the resonant frequency detunes along two furrows, but the efficiency is still high. Thus, it would be the maximum power transfer if the frequency can be tuned to the desirable frequency.

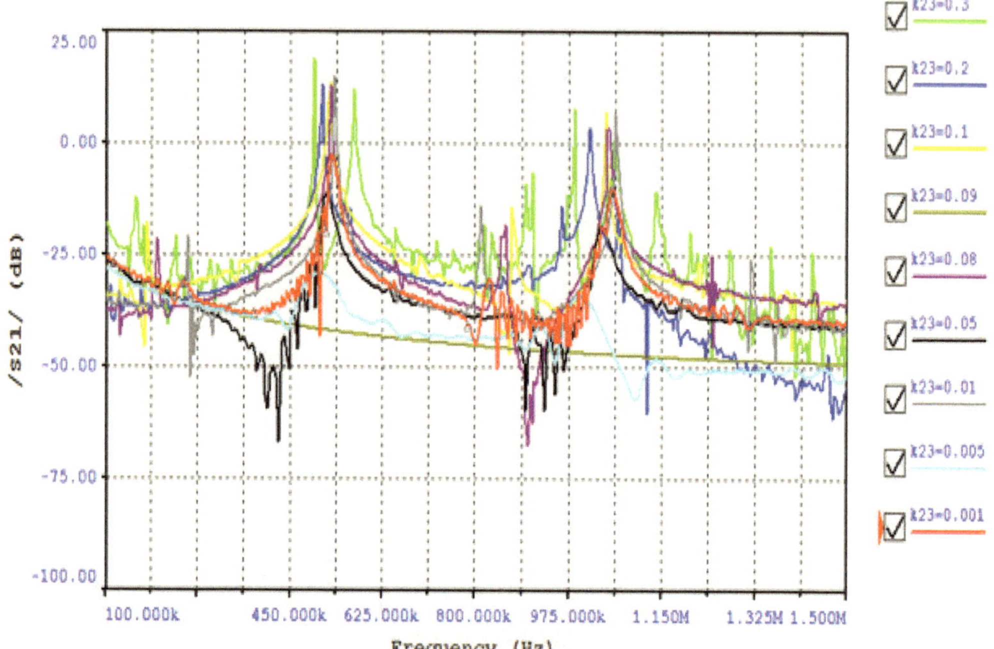

Figure 3.8. $|S_{21}|$ as a function of k_{23} and frequency

Equivalent Circuit Parameter

The two primary factors for resonant coil performance are that they resonant close to the same frequency and that they have a sufficiently high enough quality factor. The quality factor represents how well a resonant coil can hold energy without losses due to heat. The first calculation for determining the resonant frequency of a coil is to find its inductance. Equation **3.10** is reasonably accurate for calculating the inductance for a single loop inductor. Appendix A is the software used to calculate the inductance of the coil. In particular, the coil inductance

L, expressed in μH, of diameter of loop D and diameter of wire d, both expressed in mm can be written as (Wheeler, 1982):

$$L = 2 * D \left[ln \left(\tfrac{8D}{d} \right) - 1.75 \right].$$ (3.10)

It is then possible to calculate the moment frequency given capacitance and inductance with the following equation (Siskind 1980):

$$\omega_r = \frac{1}{\sqrt{L_i C_i}}, i = 1 \sim 4$$ (3.11)

where ω_r, L_i and C_i are respectively the moment frequency (Hz), equivalent capacitance (F) and inductance (μH) of each resonant circuit. The frequency of the system is simulated and it is presented in Figure 3.9.

The other component is the quality factor of the resonant loop is presented in the formula given below and computed as shown in Appendix C.

$$Q_i = \frac{1}{R_i} \sqrt{\frac{L_i}{C_i}} = \frac{\omega_r L_i}{R_i} \Leftrightarrow \omega_r L_i = R_i Q_i , \ i = 1 \sim 4$$ (3.12)

where Q_i and R_i are the quality factor of the coil and the equivalent resistance of each resonant circuit respectively (Siskind 1980).

Two ways to increase the quality factor is to lower the capacitance or resistance as seen in equation 3.12. Capacitance is easily decreased as this can be done by either pairing capacitors in series or by using new capacitors with lower value. The real effect of a lower capacitance is a higher resonant frequency as shown in equation 3.11, which limits significant currents due to voltages having less time to overcome magnetic momentum. The other method of raising quality factor, lowering resistance, can be achieved by using a lower gauge of wire for the resonant coils.

In the power coil, for instance, R_l is the sum of R_s and R_1. The resonant frequency ω_r of each coil is defined to be the same, $\omega_1 = \omega_2 = \omega_3 = \omega_4 = \omega_r$, When resonance takes place, the total impedance of each coil is presented as following;

$$Z_1 = R_s + R_1 \approx R_s \tag{3.13}$$

$$Z_2 = R_2 \tag{3.14}$$

$$Z_3 = R_3 \tag{3.15}$$

$$Z_4 = R_L + R_1 \approx R_L \tag{3.16}$$

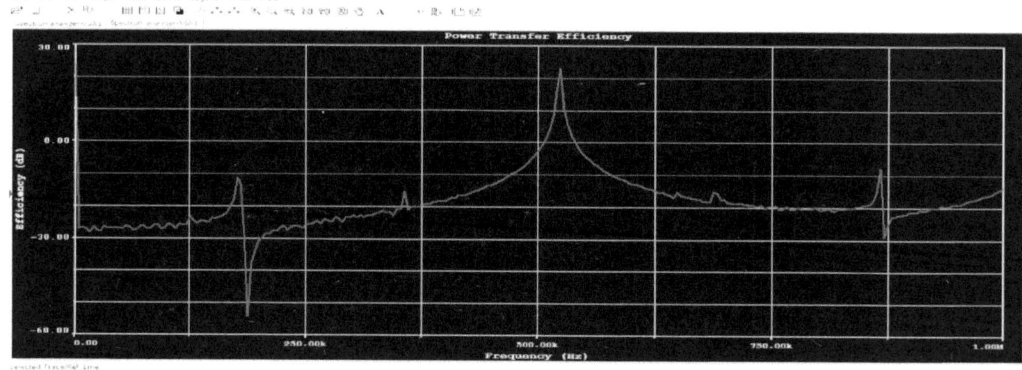

Figure 3.9. Frequency characteristics of the Transmission specification

Experimental Results and Validation

According to the structure proposed before, the experimental device has been made and study of energy transfer shows a broad application prospect for this technology in the future. Figure 3.10 showing the structure of the inductive coupling between the source (first loop on the left), the transmitting resonator (second loop from the left), receiving resonator (third loop from the left) and load (last loop on the right). Figure 3.10 shows the experimentally setup used to validate the theoretical model. The transmitter on the left consists of a small drive loop centered within a bigger loop resonator with 0.7 cm apart. The drive loop is 30 cm in diameter, with a parallel connected capacitor used to tune the system to 593.383 kHz. The large transmit loop has an outer diameter of 40 cm. The resonant frequency of 593.383 kHz was determined experimentally. The receiver is constructed similarly with the same distance as the transmitter. All elements are made of 13 mm diameter copper tube, supported by Plexiglas armatures.

One of the significant challenges when comparing the theoretical model to measured data is the accurate estimation of the lumped circuit parameters L, C and R of the physical system. To accomplish this task, we used standard RF and microwave measurement techniques developed to extract parameters such as resonant frequency, coupling coefficient, and unloaded Q factor from resonant structures.

The advantage of this technology is energy transfer can go through various objects. Different kinds of obstacles were placed between the transmitting and receiving loops respectively to test the ability of going through objects. Experimental results show that energy can still be transferred even the receiver is sheltered in most conditions. Non-metallic objects such as walls, books, wooden products, organic glass panels, leather, and textiles have no impact on power transfer.as shown in Appendix H.

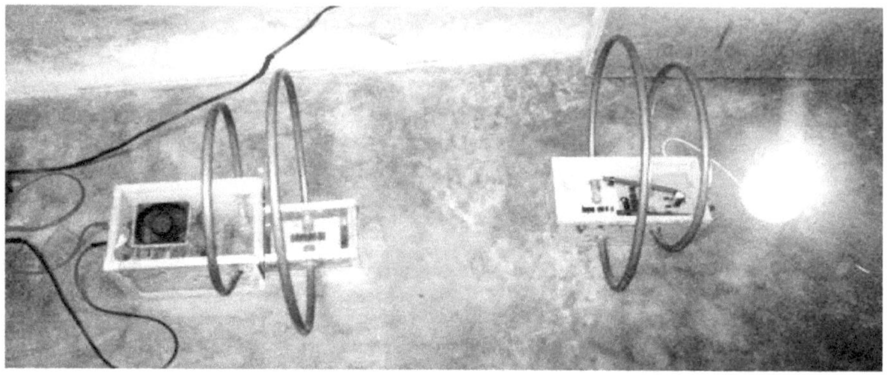

Figure 3.10. The Energy Transfer Experimental Device

The impact of metallic objects on the system depends on different characteristics of metal conductor. It would have slight impact if the object with size less than the diameter of coil as discussed earlier, or which cannot generate a larger eddy current (Zhu et al., 2008). If the metallic objects which can generate larger eddy current or form a close loop is close to this system, the impact will be greater even block energy transfer.

Based upon the data collected on Table 3.2, the following graph shows voltage as a function of distance between the loop, the transmitter and receiver. Curves in figure 3.11 show the impact on the output voltage caused by different transfer distance. It is obvious that the longer is the distance between the two loops, the lower the output voltage of the receiving loop is. Since the coefficient of determination R^2 has a value close to 1 for the linearly fit, the data points were strongly correlated.

Table 3.2. Relationship between Receiver Output Voltage and Distance

Distance (cm)	Output Voltage (Volts)	Distance (cm)	Output Voltage (Volts)	Distance (cm)	Output Voltage (Volts)	Distance (cm)	Output Voltage (Volts)
0	10.83	30	8.09	60	2.73	90	0.603
5	10.15	35	7.04	65	2.07	95	0.469
10	9.79	40	6.29	70	1.57	100	0.345
15	9.68	45	5.43	75	1.212	105	0.246
20	9.45	50	4.83	80	0.975	110	0.006
25	9.24	55	3.58	85	0.742	115	0.007

In other words, beyond the coupled mode region, the Power Transfer Efficiency decreases rapidly as a function of distance. This illustrates the theory when the distance between the two antennas increases, the coupled mode resonance phenomenon disappears, and the antennas behave like traditional transmitting and receiving antennas in the near field.

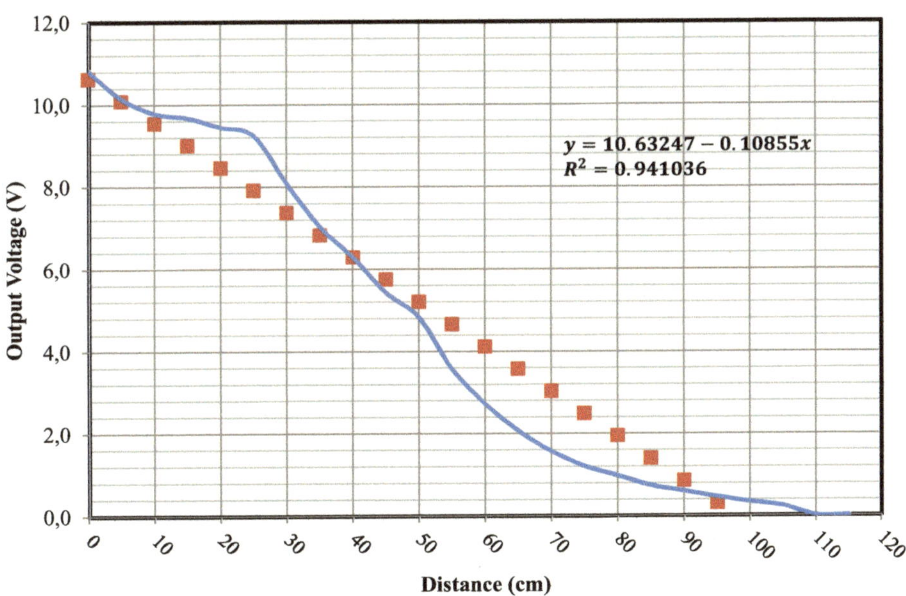

Figure 3.11. Relationship between the transfer distance and output voltage of the load loop

Also, the graph below shows that the voltages gain as a function of distance, the lesser the distance, the greater in the gain voltage. Hence, greater distance between the transmitting and the receiving coils given lower gain voltage. It is further observed from the graph there is a limiting distance of the two coils.

The statistic multiple R suggests that efficiency is highly related to the distance between the coils. Moreover, the R-square (R^2) suggests that 93.32 percent of the variance in the efficiency tested can be explained by the distance of the two coils during the experiment. The data of this graph depicted in Appendix F. This result could be interpreted to mean that there is enough evidence to show that distance between the coils determines higher gain voltage of the system.

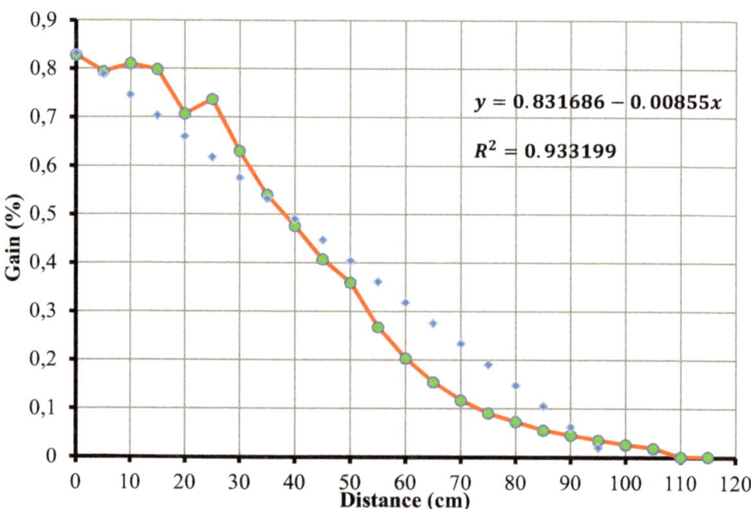

Figure 3.12. Relationship between the Voltage Gain and the Separation Distance

System Efficiency

The power transfer efficiency, which describes the direct energy transfer between the transmitter and receiver coils, as discussed in equation 3.8, is measured as the ratio of the power received in the receiver coil to the power sent out from the transmitter coil. The table and graph below shows that the lesser the distance, the greater in the efficiency. Hence, greater distance between the transmitting and the receiving coils given lower efficiency. It is further observed from the graph there is a limiting distance of the two coils. This is the distance where the graph crossed the zero efficiency line.

Table 3.3. Measured Value for Voltage, Current and Power

Distance (cm)	Input			Output		
	Voltage (Volts)	Current (Ampere)	Power (Watts)	Voltage (Volts)	Current (Ampere)	Power (Watts)
0	5.61	7.72	43.3092	8.5	2.05	17.425
5	4.26	8.75	37.275	7.29	1.95	14.2155
10	11.48	2.75	31.57	4.54	1.6	7.264
15	12.63	1.94	24.5022	2.28	1.31	2.9868
20	12.68	1.85	23.458	1.79	0.52	0.9308
25	13.09	1.79	23.4311	1.72	0.31	0.5332
30	13.01	1.75	22.7675	1.702	0.18	0.30636
35	13.29	1.72	22.8588	1.67	0.11	0.1837
40	13.18	1.72	22.6696	1.645	0.07	0.11515
45	13.3	1.69	22.477	1.625	0.04	0.065
50	13.31	1.71	22.7601	1.59	0.02	0.0318
55	13.35	1.69	22.5615	1.5	0.01	0.015

The statistic multiple R suggests that efficiency is highly related to the distance between the coils. Moreover, the R-square (R^2) suggests that 71.15 percent of the variance in the efficiency tested can be explained by the distance of the two coils during the experiment.

Moreover, the analysis of variance yielded an F=24.66 which is significant even beyond the 0.01 level. Actually, the significance level is 0.000564872. This result could be interpreted to mean that there is enough evidence to show that distance between the coils determines higher efficiency of the prototype.

This result finds support from the studies of (Sample et al., 2009), (Kurs et al., 2007), (Park and Kim, 2012) and (Yoon and Ling, 2012).

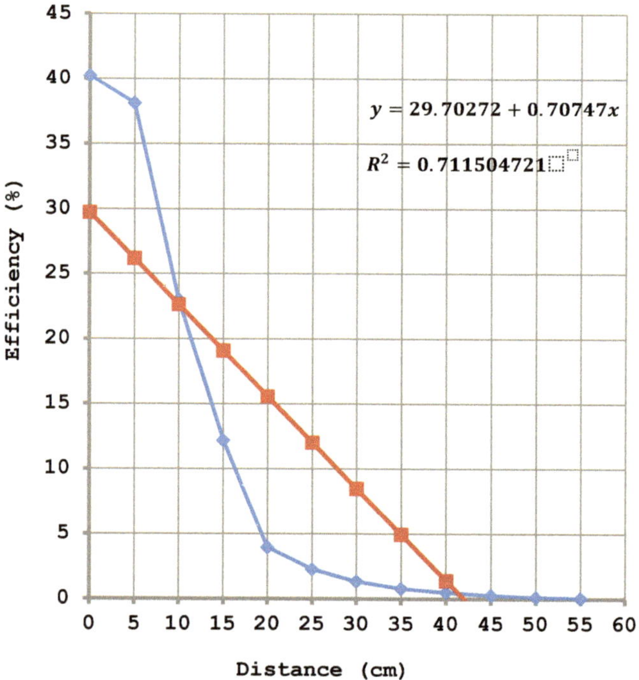

Figure 3.13 . Relationship between the Transfer Distance and System Efficiency

Voltage Patterns as a Function of Angular Displacements

In this experiment the generating coil was kept in a fixed position while the receiving coil take samples of the electromagnetic field, around the generating coil, at a fixed distance and with constant angular displacement completing 360 degrees as shown in Figure 3.12.

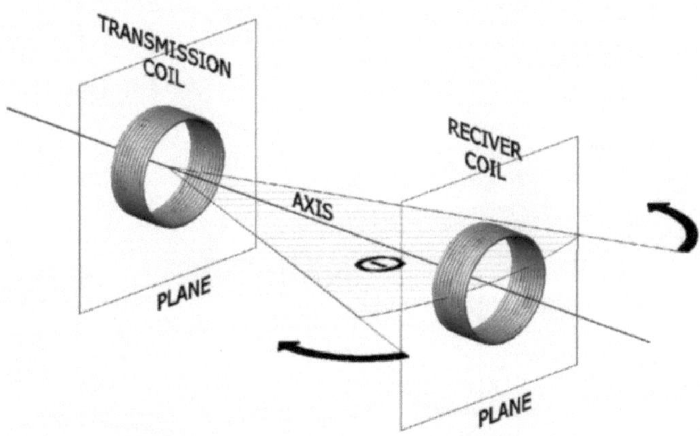

Figure 3.12. Experimental process for the revolving coil (Herrera, 2010)

On the following graph, figure 3.13 the radiation patterns for the generating coil is shown. Also, Appendix E is the tabulated data of the pattern. From this experiment can be seen that the produced energy by the generating coil spreads at **90°** in front of the generating coil and at **90°** behind the same coil. This result becomes interesting because the next experiment will be designed taking into account the radiation pattern in order to send the highest forward energy and take account of the rear for further experiments. It can also be seen that the highest radiation level is located at **345°** and at **165°** (due to the configuration

26

and experiment conditions, the receiving coil had a displacement of about **15**° to the x-axis),
that is, just in front and back of the generating coil. The fact to have a bidirectional pattern
has an important effect on the gain calculation of the system, given the fact that energy
depends on the directivity of the coil. Thus, the efficiency will be affected by the shape of the
radiation pattern.

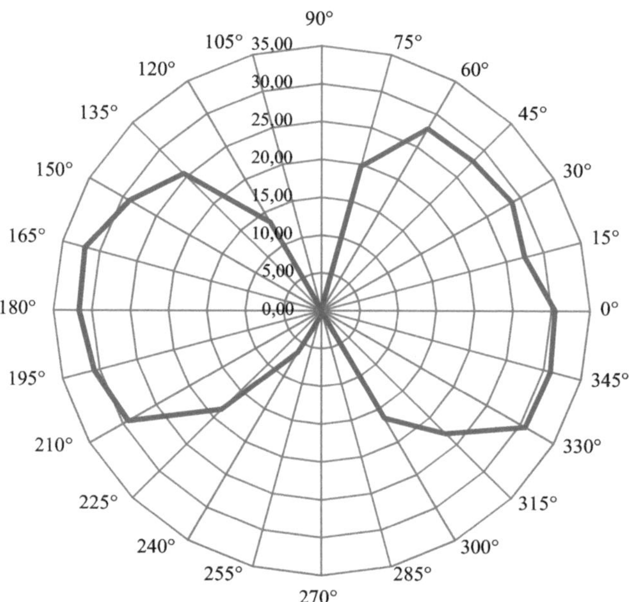

Figure 3.13. Voltage pattern in different orientation

To measure the voltage gain both loops were placed facing each other and the receiving
loop was moved away at a constant rate from the generating loop. Since the generating loop
has constant voltage, the voltage measurement was only performed at the receiving loop for
each displacement.

Once the voltages were measured, the ratio between them was graph and is shown in figure 3.14. In this graph can be seen that when both loops are close together all the energy is sent to the front by the generating loop which is taken by the receiving loop. This is noticed when the voltage gain is 50%. The result is logical after observing the radiation pattern shown in Figure 3.13, because a radiation back lobe is wasted. Figure 3.14 shows, beyond the 8 cm. distance, the voltage gain for the system falls below the 5% value. The back lobe could be reused using a reflecting surface for the magnetic field.

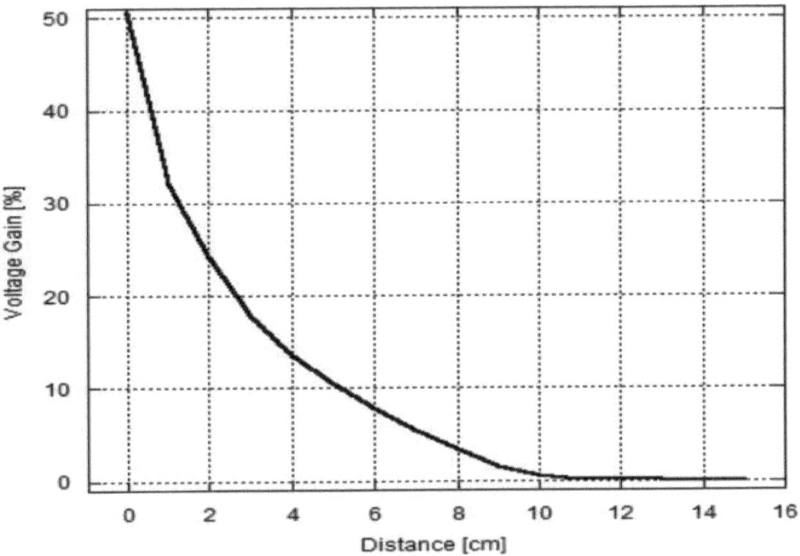

Figure 3.14. Voltage gain of the system with respect to the voltage ratio

Testing at Different Frequencies

Figure 3.15 shows that the relationship between the different driving frequencies, the resonant frequency and the back emf's on the load loop. As depicted on the graph that driving frequencies have slight impact on the amplitude side. On the side of resonant frequency it seen that there is no effect on the part of driving different frequencies.

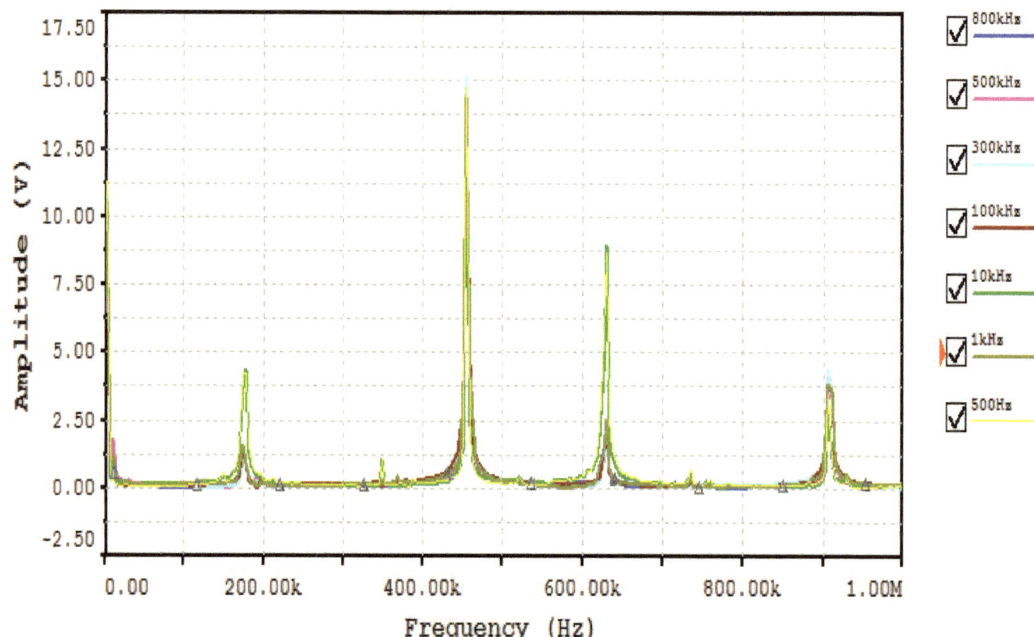

Figure 3.15. Relationship Between the Different driving Frequency, Resonant Frequency and the Output Voltage on Load Loop

Conclusion

Wireless energy transfer is completely has potential to be improved and magnetic resonance coupling is most feasible method in wireless energy transfer. Otherwise, this paper also proved that the electrical energy transmission will transpire the most efficient in the resonance frequency. The application was used to prove this by using cell phone, dc motor, led and bulb. In addition, it can conclude that the maximum output voltage at receiver of the resonant wireless energy transfer depends on the size of coil and the input voltage. The transfer energy could be done effectively by increasing the input voltage. The size and frequency of the device also should be taken into account since the resonance wireless energy transfer device works in the medium or high frequency range of electromagnetic field. The higher the frequency, the closer is the common coil and capacitor to the resonant condition.

This work specifically lays down the ground work of innovative wireless power technology and open opportunities to commercially implement advanced electromagnetic resonance based on WPT systems. Based on the experimental result and analysis, it is proven that energy transfer is completely feasible using the technology of magnetic coupling resonance.

Simple and common processes based on the electrical circuit theory are applied to the analyses throughout this paper in order to obtain the initial and deterministic design parameters of the WPT system comprising resonators and their coupling elements and achieve a pass band at the frequency of interest. Also, circuit simulations have been performed using Multisim.

Beyond the coupled mode region, the Power Transfer Efficiency decreases rapidly as a function of distance. This illustrates the theory when the distance between the two antennas increases, the coupled mode resonance phenomenon disappears, and the antennas behave like traditional transmitting and receiving antennas in the near field.

It was verified that the distance between the coils center play an important role on the efficiency of the power transfer, decreasing as the center are moved away and reaching its maximum at 0 cm. Besides, the relative angles between planes of each coil also affect the efficiency, establishing in experiment that the planes should be placed in parallel fashion over the same axis. This theoretical model is validated against measured data and shows an excellent average coefficient of determination R^2 of 0.937118 that signifies 93.7118 percent of the variation in the distance can be explained through the linear regression relation.

References

Ali M., Yang, G. and Dougal, R. (2005). A New Circularly Polarized Rectenna for Wireless Power Transmission and Data Communication. IEEE Antennas and Wireless Propagation Letters, Volume 4, pages 205-208.

Basset, P., Andreas Kaiser, B. L., Collard, D. & Buchaillot, L. (2007). Complete System for Wireless Powering and Remote Control of Electrostatic Actuators by Inductive Coupling, IEEE/ASME Transactions on Mechatronics 12(1).

Brown. W. (1984). The History of Power Transmission by Radio Waves. IEEE Transactions on Microwave Theory and Techniques Vol.MTT-32, No.9:1230-1242.

Dickinson, R. (2003). Wireless Power Transmission Technology State of the Art at First Bill Brown Lecture. Science Direct: 561-570

Dionigi, M., Costanzo, A. and Mongiardo, M. (2012). Network Methods for Analysis and Design of Resonant Wireless Power Transfer Systems, Wireless Power Transfer – Principles and Engineering Explorations, Dr. Ki Young Kim (Ed.), ISBN: 978-953-307-874-8

Eom, K. & Arai, H. (Mar. 2009). Wireless power transfer using sheet-like waveguide, in Proc. EuCAP, Berlin, Germany, pp. 3038–3041

Fotopoulou, K. & Flynn, B. W. (Feb. 2011). Wireless power transfer in loosely coupled links: coil misalignment model, IEEE Trans. Magn., vol. 47, no. 2, pp. 416–430

Gao, J. (2007). Traveling Magnetic Field for Homogeneous Wireless Power Transmission, IEEE Transactions on Power Delivery 22(1)

Glaser, P. E. (1973). Method and apparatus for converting solar radiation to electrical power, U.S.A Patent

Grajski, K., Tseng, R. and Wheatley, C. (2012). Loosely-Coupled Wireless Power Transfer: Physics, Circuits, Standards. Microwave Workshop Series on Innovative Wireless Power Transmission: Technologies, Systems and Applications (IMWS), 2012, IEEE MTT-S International. DOI: 10.1109/IMWS.2012.6215828

Hansen, J. E. (1988). Spherical near-field antenna measurements, IEE Electromagnetic Waves Series 26

Herrera, J.A., Torres, H., Leal, H. and Angel, A. (2010). Experiment about wireless energy transfer, International Congress on instrumentation and Applied Sciences, CCADET, Cancun, Q.R.,Mexico, pp. 1–10.

Hirayama, H. (2012). Equivalent Circuit and Calculation of Its Parameters of Magnetic Coupled-Resonant Wireless Power Transfer. Wireless Power Transfer - Principles and Engineering Explorations, Dr. Ki Young Kim (Ed.), ISBN: 978-953-307-874-8

Hoang H. and Bien, F. (2012). Maximizing Efficiency of Electromagnetic Resonance Wireless Power Transmission Systems with Adaptive Circuits. Wireless Power Transfer-Principles and Engineering Explorations, Dr. Ki Young Kim, ISBN: 978-953-307-874-8

Imura, T., Okabe, H. & Hori, Y. (Sep. 2009). Basic experimental study on helical antennas of wireless power transfer for Electric Vehicles by using magnetic resonant couplings, in Proc. IEEE VPPC, Dearborn, MI, pp. 936–940

Ishiyama, T., Kanai, Y., Ohwaki, J. and Mino, M. (2003) Impact of Wireless Power Transmission system Using Ultrasonic air Transducer for Low Power   Mobile Applications. IEEE Ultrasonics Sympossium: 1368-1371

Jang, B.J.,Lee, S. and Yoon, H. (2012). HF-Band Wireless Power Transfer System: Concept, Issues, and Design. Progress In Electromagnetics Research, Vol. 124, page 211-231

Kim, H.S., Wono, D.H. and Jang, B.J. (2010). Simple design method of wireless power transfer system using 13.56MHz loop antennas. 978-1-4244-6392-3

Komaru, T., Koizumi,M., Komurasaki, K., Shibata, T. and Kano, K. (2012). Compact and Tunable Transmitter and Receiver for Magnetic Resonance Power Transmission to Mobile Objects, Wireless Power Transfer - Principles and Engineering Explorations, Dr. Ki Young Kim (Ed.), ISBN: 978-953-307-874-8

Kurs, A., Karalis, A., Moffatt, R., Joannopoulos, J. D., Fisher, P. and Soljacic, M., (2007), Wireless Power Transfer via Strongly Coupled Magnetic Resonances. Science Direct, Vol. 317, 83-86

Kurs, A., Karalis, Joannopoulos, J. D., and Soljacic, M., (2008), Efficient wireless non-radiative mid-range energy transfer. Science Direct, Annals of Physics 323 (2008) 34–48

Lee, J. & Nam, S. (Nov. 2010). Fundamental aspects of near-field coupling small antennas for wireless power transfer, IEEE Trans. Antennas Propag., vol. 58, no. 11, pp.3442–3449

Low, Z. N., Chinga, R. A., Tseng, R. and Lin, J. (2009). Design and Test of a High-Power High-efficiency Loosely Coupled Planar Wireless Power Transfer System, IEEE Transactions on Industrial Electronics 56(5)

Mansor, H., Halim, M., Mashor, M. and Rahim, M. (2008). Application on Wireless Power Transmission for Biomedical Implantable Organ, Springer-Verlag Biomed 2008 proceedings 21 pp. 40–43

Mokalkar, B., Tale, C. and Edle J. (2012). Witricity: A Novel Concept of Power Transfer, International Journal of Engineering Inventions. ISSN: 2278-7461, www.ijeijournal.com Volume 1, Issue 7 (October2012) PP: 51-59

NASA (2003). Beamed Laser Power for UAVs, Dryden Flight Research Center

Park,Y., Kim, J. and Kim, K.H. (2012). Magnetically Coupled Resonance Wireless Power Transfer (MR-WPT) with Multiple Self-Resonators, Wireless Power Transfer - Principles and Engineering Explorations, Dr. Ki Young Kim (Ed.), ISBN: 978-953-307-874-8

Poon, A. S. Y., O'Driscoll, S. & Meng, T. H. (May 2010). Optimal frequency for wireless power transmission into dispersive tissue, IEEE Trans. Antennas Propag., vol. 58, no. 5, pp. 1739–1750

RamRakhyani, A. K., Mirabbasi, S. & Chiao, M. (Feb. 2011). Design and optimization of resonance-based efficient wireless power delivery systems for biomedical implants, IEEE Trans. Biomed. Circuits Syst., vol. 5, no. 1, pp. 48–63

Ren, Y.-J. and Chang, K. (2006). 5.8-ghz Circularly Polarized Dual-Diode Rectenna and Rectenna Array for Microwave Power Transmission, IEEE Transactions on Microwave Theory and Techniques 54(4)

Sample, A., Meyer, D. and Smith, J. (2009). Analysis, Experimental Results, and Range Adaptation of Magnetically Coupled Resonators for Wireless Power Transfer, IEEE Explore. Intel Corporation via the Intel Library

Schantz, H. G. (Jul. 2005). A near field propagation law & a novel fundamental limit to antenna gain versus size, in IEEE Antennas Propag. Int. Symp. Dig., Washington, DC

Sedwick, R. (2012). A Fully Analytic Treatment of Resonant Inductive Coupling in the Far Field, Wireless Power Transfer - Principles and Engineering Explorations, Dr. Ki Young Kim (Ed.), ISBN: 978-953- 307-874-8

Shams, K. M. Z. & Ali, M. (2007). Wireless Power Transmission to a Buried Sensor in Concrete, IEEE Sensors Journal 7(12)

Siskind, C. (1980). Electrical Circuits, Second Edition, Philippine Copyrigh McGraw-Hill, Inc.

Stielau, O.H. and Covic, G.A. (2000). Design of Loosely Coupled Inductive Power Transfer Systems, IEEE: 35-90

Sugiyama, H. (2012). Performance Analysis of Magnetic Resonant System Based on Electrical Circuit Theory, Wireless Power Transfer - Principles and Engineering Explorations, Dr. Ki Young Kim (Ed.), ISBN: 978-953-307-874-8

Thompson, M. (1999). Inductance Calculation Techniques—Part II: Approximations and Handbook Methods. Power Control and Intelligent Motion

Trezise, T. (2011), Modelling Inductively Coupled Coils for Wireless Implantable Bio Sensors, A Novel Approach Using the Finite Element Method, University of Victoria

Vandevoorde, G. and Puers, R. (2001). Wireless Energy Transfer for Stand Alone Systems: A Comparison between Low and High Power Applicability. Sensor and Actuators: 305-311.

Yaghjian, A. D. (Jan. 1982). Efficient computation of antenna coupling and fields within the near-field region, IEEE Trans. Antennas Propag., vol. 30, no. 1, pp. 113–128

Yoon, I.J. and Ling, H. (2012). Realizing Efficient Wireless Power Transfer in the Near Field Region Using Electrically Small Antennas, Wireless Power Transfer - Principles and Engineering Explorations, Dr. Ki Young Kim (Ed.), ISBN: 978-953-307-874-8

Yu, X., Herraul, F., Ji, C.H., Seong-Hyok Kim, S. K., Allen, M., Lisi, G., Nguyen, L. and Anderson, D. (2011). Watt-Level Wireless Power Transfer Based on Stacked Flex Circuit Technology. IEEE Electronic Components and Technology Conference , 978-1-61284-498-5

Zhu, C., Liu, K., Yu,C., Ma R. and, Cheng, H. (2008). Simulation and Experimental Analysis on Wireless Energy Transfer Based on Magnetic Resonances. IEEE Vehicle Power and Propulsion Conference (VPPC), Harbin, China

YOUR KNOWLEDGE HAS VALUE

- We will publish your bachelor's and
  master's thesis, essays and papers

- Your own eBook and book -
  sold worldwide in all relevant shops

- Earn money with each sale

Upload your text at www.GRIN.com
and publish for free